my guide to the planets

Mars

Cherry Lake Press

Published in the United States of America by Cherry Lake Publishing
Ann Arbor, Michigan
www.cherrylakepublishing.com

Reading Adviser: Marla Conn, MS, Ed, Literacy specialist, Read-Ability, Inc.
Book Designer: Jennifer Wahi
Illustrator: Jeff Bane

Photo Credits: ©Vadim Sadovski/Shutterstock, 5, 7, 9, 15, 19; ©Triff/Shutterstock, 11; ©Jurik Peter/Shutterstock, 13; ©Semiglass/Shutterstock, 17; ©esfera/Shutterstock, 21; ©NASA images/Shutterstock, 23; Cover, 2-3, 6, 8, 22, Jeff Bane; Various vector images throughout courtesy of Shutterstock.com/

Library of Congress Cataloging-in-Publication Data

Names: Devera, Czeena, author. | Bane, Jeff, 1957- illustrator. | Devera, Czeena. My guide to the planets.
Title: Mars / by Czeena Devera ; illustrated by Jeff Bane.
Description: Ann Arbor, Michigan : Cherry Lake Publishing, [2020] | Series: My guide to the planets | Includes index. | Audience: K-1.
Identifiers: LCCN 2019032853 | ISBN 9781534158825 (hardcover) | ISBN 9781534161122 (paperback) | ISBN 9781534159976 (adobe pdf) | ISBN 9781534162273 (ebook)
Subjects: LCSH: Mars (Planet)--Juvenile literature.
Classification: LCC QB641 .D48 2020 | DDC 523.43--dc23
LC record available at https://lccn.loc.gov/2019032853

Printed in the United States of America
Corporate Graphics

table of contents

About the author: Czeena Devera grew up in the red-hot heat of Arizona surrounded by books. Her childhood bedroom had built-in bookshelves that were always full. She now lives in Michigan with an even bigger library of books.

About the illustrator: Jeff Bane and his two business partners own a studio along the American River in Folsom, California, home of the 1849 Gold Rush. When Jeff's not sketching or illustrating for clients, he's either swimming or kayaking in the river to relax.

About Mars

I'm Mars. I am the fourth-closest planet to the Sun.

I am the second-smallest planet.

I **orbit** around the Sun. It takes me about 687 days to complete 1 orbit.

Gravity is a force. It keeps things from floating away. I have less gravity than Earth. This means you weigh more on Earth than on me.

I’m called the red planet. My rocks are red from rust. I even look red from Earth!

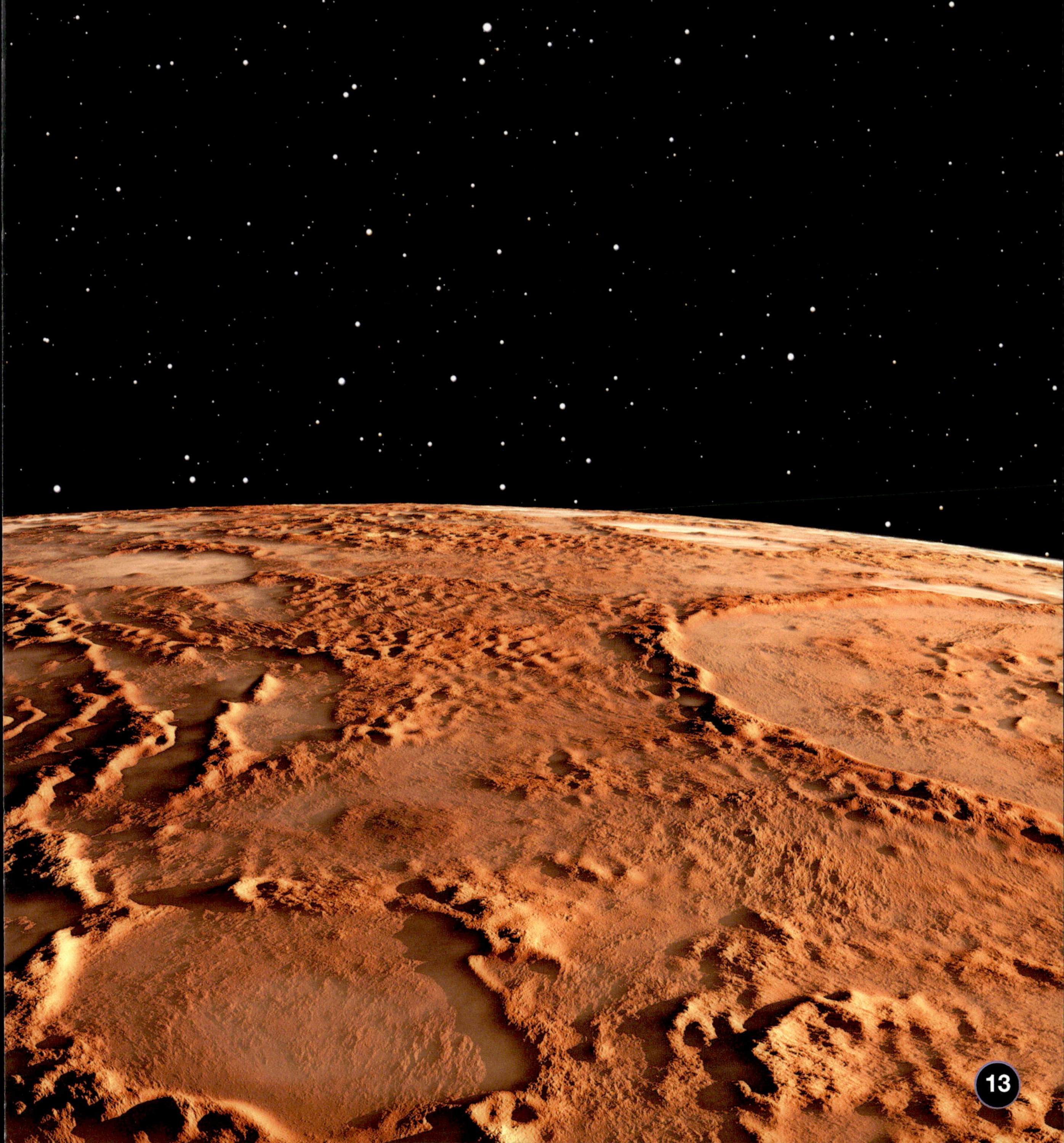

I have two moons. **Scientists** believe these moons were once **asteroids**.

I have a cold and thin **atmosphere**. This means liquid water freezes almost instantly.

I have the largest mountain in the **solar system**. It's a volcano.

I also have cliffs and valleys.

I am a **unique** planet. Scientists believe I can support life one day!

glossary & index

glossary

asteroids (AS-tuh-roidz) small, rocky objects that travel around the sun

atmosphere (AT-muhs-feer) the mixture of gases that surrounds a planet

orbit (OR-bit) to travel in a curved path around something

scientists (SYE-uhn-tists) people who study nature and the world we live in

solar system (SOH-lur SIS-tuhm) the Sun and all the things that orbit around it, like planets

unique (yoo-NEEK) the only one of its kind

index